Gabriel Pereira de Macedo
Carlos Alberto

NDT techniques for detecting flaws in plates in naval structures

AF301662

Gabriel Pereira de Macedo
Carlos Alberto

NDT techniques for detecting flaws in plates in naval structures

Identification of flaws in carbon steel welded joints

ScienciaScripts

Imprint

Any brand names and product names mentioned in this book are subject to trademark, brand or patent protection and are trademarks or registered trademarks of their respective holders. The use of brand names, product names, common names, trade names, product descriptions etc. even without a particular marking in this work is in no way to be construed to mean that such names may be regarded as unrestricted in respect of trademark and brand protection legislation and could thus be used by anyone.

Cover image: www.ingimage.com

This book is a translation from the original published under ISBN 978-613-9-62084-5.

Publisher:
Sciencia Scripts
is a trademark of
Dodo Books Indian Ocean Ltd. and OmniScriptum S.R.L publishing group

120 High Road, East Finchley, London, N2 9ED, United Kingdom
Str. Armeneasca 28/1, office 1, Chisinau MD-2012, Republic of Moldova, Europe
Printed at: see last page
ISBN: 978-620-7-71241-0

Copyright © Gabriel Pereira de Macedo, Carlos Alberto
Copyright © 2024 Dodo Books Indian Ocean Ltd. and OmniScriptum S.R.L publishing group

SUMMARY

ACKNOWLEDGMENTS

To UNESA, for providing me with excellent facilities to qualify as a Petroleum Engineer and for the good knowledge I acquired at the University.

To the teachers I had throughout the course, for sharing their theoretical and practical knowledge in order to train us for the job market.

To Professor Carlos Alberto for agreeing to be my advisor on this project and helping me a lot professionally.

I would also like to thank the course coordinator for making this work serious and compatible with the job market.

To my colleagues on the Petroleum Engineering course, for the camaraderie and the knowledge acquired through the debates we had.

I would also like to thank the team at PWD END, SPERJ and VARD NITERÓI S.A., companies that helped carry out the experiments presented in this work.

SUMMARY

This work is a study carried out at the VARD Niterói S.A. shipyard in the area of non-destructive testing, using the methods of liquid penetrant (LP), ultrasound (US) and industrial radiography (IR), in order to detect discontinuities in steel plates and welding processes to maintain the quality of the material and where it will be used.

Keywords: Non-destructive testing, NDT, Liquid Penetrant, X-ray, Ultrasound.

1   INTRODUCTION

The art of inspecting without destroying has evolved greatly, especially since the 1950s, and today it is an indispensable production tool. Today, non-destructive testing is widely used in modern industry all over the world to assess quality, detect small surface flaws, the presence of cracks and other physical interruptions such as pores, determine material thicknesses and coatings, among other discontinuities.

Non-destructive tests (NDTs) are those which, when carried out on finished or semi-finished parts, do not leave any marks or signs on the parts or render them unusable, unlike destructive tests which leave signs on the part or even render it unusable.

Discontinuities are interruptions in the structure of a material, at a macro or microscopic level, which can be seen during a test, among which we highlight X-ray, Liquid Penetrant and Ultrasound tests, which we will talk about more specifically as they are the most widely used in industries today to certify the reliability and quality control of the product being manufactured. Tests on the joints of welded parts, during and after the welding process, will be highlighted.

In the tests carried out, standards, procedures and qualified professionals were used in order to allow comparison between the various results and to establish levels of requirements for each test.

Among the main advantages of non-destructive testing are the prevention of accidents, where a crack or a pore in the weld at the junction of parts can cause a break and consequently an accident in the structure, other advantages that can be cited are the guarantee of customer satisfaction, cost reduction, since in non-destructive testing their costs are greatly reduced and maintenance of product quality.

2 BACKGROUND

The economic losses caused by the corrosive process of equipment deterioration are enormous. That's why it's never too much to think about techniques or processes that can predict their occurrence in a timeframe that allows decisions to be taken before the failures are catastrophic and the damage is large-scale.

Also due to the aforementioned fact, in recent years there has been a significant increase in the application of non-destructive testing in industrial equipment based on the following facts:

> They can be used as a cost-effective tool in predictive maintenance;

> They can provide important information for establishing the structural integrity of equipment;

> They make it possible to apply life extension techniques based on the planned replacement of deteriorated components.

This trend implies the need for some modifications to the inspection system normally adopted when using conventional non-destructive testing:

> Total inspection of equipment instead of sample testing;

> Inspection carried out while the equipment is in service (operation). This attitude implies the use of non-destructive tests on hot surfaces and/or the execution of tests on insulation; these facts make inspection almost prohibitive for conventional non-destructive tests, as they usually do not allow a quick definition of the critical regions of the equipment.

3 OBJECTIVES

This paper presents X-ray, Liquid Penetrant and Ultrasound tests carried out at the Vard Niterói S.A. shipyard in conjunction with PW END and SPERJ, companies specialized and certified to carry out these tests, with the aim of verifying the effectiveness of a procedure that uses equipment commonly used in conventional inspection to detect and locate discontinuities in welds in carbon steel plates and pipelines.

Non-destructive testing is one of the main tools for controlling product quality and is widely used in the oil, petrochemical and shipbuilding industries, among others. Their main characteristics are that they prioritize the quality of goods and services, reduce costs, preserve the environment and are the main competitive factor among the companies that use them.

The test methods carried out are capable of providing information about the defect content of a given product, the characteristics of a material, which can reveal and monitor the degradation of components, equipment and structures.

This shows that simple, low-investment procedures guarantee faults in manufacturing processes with great precision, thus demonstrating the quality and reliability of products in their manufacturing phase, generating reports as future guarantees for manufactured products.

4 THEORETICAL BASIS

4.1 NON-DESTRUCTIVE TESTING (END)

Non-destructive testing (NDT) is testing carried out on finished or semi-finished materials to verify the existence or not of discontinuities or defects using defined physical principles, without altering their physical, chemical, mechanical or dimensional characteristics and without interfering with their subsequent use.

NDT methods include those capable of providing information about the defect content of a product. The methods we used for experimental tests were X-rays, liquid penetrant and ultrasound to detect discontinuities in welds.

The main discontinuities in welds are cracks in the final crater of the bead, porosity, slag or tungsten inclusions, lack of penetration, lack of lateral fusion and bites.

Cracks are generated as a result of the thermal stresses generated during welding and the material's inability to deform to absorb this stress.

Porosities are caused by gases that were unable to escape during the solidification of the weld.

Inclusions are usually the result of insufficient cleaning between welding passes or incorrect handling of the wire during the operation.

Lack of penetration or fusion is caused by a lack of sufficient energy to promote fusion of the joint. This can occur due to high welding speed, incorrect handling of the equipment, among other causes.

Bites are caused by high welding speeds or very long arc lengths.

4.1.1 X-RAY TESTING

According to ANDREUCCI, when you want to inspect parts in order to detect internal defects, radiography is a powerful method for detecting discontinuities a few millimeters long with high sensitivity. Radiography is a method mainly used in the oil and petrochemical industries, power generation for inspecting mainly welds, they play an important role in proving the quality of the part or components in compliance with the requirements of standards, specifications, among others.

Industrial radiography is essential and in some ways unsurpassed in documenting the quality of the product being inspected, because the image projected from the radiographic film represents an internal photograph of the part, which no other non-destructive test can show. This is why you need a specialized technician and to follow the test procedures.

Radiography is a method based on the differentiated absorption of penetrating radiation by the part being inspected. With variations in thickness and differences in density, there can be a difference in the absorption of penetrating radiation. This difference can be detected using film, an image tube or electronic radiation detectors. This variation in the amount of absorbed radiation detected will indicate the existence of an internal flaw or defect in the material. It is capable of detecting volumetric defects with good sensitivity, i.e. defects with small thicknesses such as voids and inclusions will be easily detected as long as they are not too small in relation to the thickness of the part.

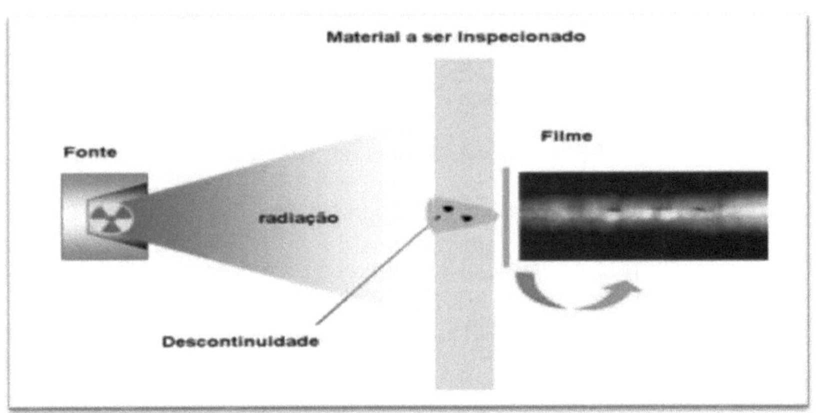

Figure 1 - X-ray testing process (ANDREUUCCI, 2008)

This test allows a permanent record to be made since a film is used and all the evidence of the inspection remains on it in an objective and incontrovertible way, compared to other methods which are to be analyzed by the inspector at the time of the test. Gamma-ray testing can be carried out in the field, i.e. on open installations such as pipelines, storage tanks, blast furnaces, among many others. Another advantage of radiography is that it is relatively inexpensive.

The disadvantages of the test, however, are the ionizing radiation that sensitizes the film, from both X-rays and gamma rays, which are highly harmful to human health. Professionals must be properly qualified and have extensive knowledge of the manufacturing and welding processes of the materials being tested in order to be able to grade the radiographic film.

X-rays are produced in special glass ampoules, the size of which can vary according to the operating voltage of the device. Special attention must be paid to the target, contained in the anode. Its surface is thus hit by an electronic flux called a thermal focus, coming from the filament.

X-ray equipment is mainly divided into two types:

> Control panel

The control panel is equipped with a box housing all the controls, indicators, switches and meters, as well as containing all the high-voltage generating circuit equipment. It is through the control panel that the voltage and amperage settings are made, as well as the device's drive command.

> Head

The head houses the bulb and the cooling device, which is connected to the control panel via special high-voltage cables.

The exposure or amount of radiation received is directly proportional to the milliamperage and the voltage applied. Once these two factors are fixed, the exposure time is the parameter that influences the sensitization of the film, i.e. the longer the time, the greater the sensitization. To determine the exposure time, the exposure curve supplied by the equipment manufacturer is used.

The main characteristics of radiographic films are: radiographic density, image contrast, film speed and grain size and they are classified according to table 1.

Table 1 Characteristics of Industrial Radiology films

TYPE OF FILM	SPEED	CONTRAST	GRANULATION
1	low	very high	extra fine
2	average	high	thin
3	high	medium	coarse

| 4 | very high | very high | several |

Source: Materials Testing - Essel

The selection of radiographic film for a given application is a compromise between the quantity of radiography required and its cost, including exposure time. To make the choice easier, manufacturers provide characteristic curves for each type of film.

The **American Society for Testing and Materials - ASTM E94** standard relates the type of film to the thickness of the part and the voltage to be developed in the test. Table 2 shows an extract from this standard developed for steel.

Table 2 ASTM E94 standard for radiographic film selection

FILM SELECTION GUIDE ACCORDING TO ASTM E94 - FOR STEEL

MATERIAL THICKNESS (mm)	VOLTAGE APPLIED TO THE EQUIPMENT (KV)				
	50 a 80	80 a 120	120 a 150	150 a 250	250 a 400
0 a 63	3	3	3	1	
6,3 a 12,7	4	3	2	2	1
12,7 a 25,4		4	3	2	2
25,4 a 50,8				3	2
50,8 a 101,6				4	4
101,6 a 203,2					4

Source: Materials Testing - Essel

4.1.2 LIQUID PENETRANT TEST

Liquid penetrant testing according to PUJATTI is characterized by the ease with which the method can be applied, i.e. it can be applied in industrial facilities, workshops or in the field, regardless of the availability of resources, which other methods require. The advantages are low cost, ease of application, high sensitivity and the possibility of using it on materials such as iron, steel, aluminum, titanium or nickel alloys, ceramics, glass, in manufacturing processes such as casting, forging, rolling, welding and in checking materials in service for fatigue or stress corrosion. The test is normally applied to surfaces at room temperature.

In Liquid Penetrant testing, it is possible to detect surface discontinuities such as cracks, pores, folds, etc. The testing process is characterized by the basic use of two products: the Penetrant and the Developer.

The penetrant is used to penetrate open surface discontinuities and form indications, the main property of which is its ability to penetrate fine openings. For a good penetrant to be functional and of good quality, it must have the following properties:

> Viscosity

A higher viscosity takes longer to enter a discontinuity, but a lower viscosity tends not to remain on the surface of the part for long, which can lead to insufficient time for penetration.

> Surface tension

The surface tension of a liquid is the result of the cohesive forces between the molecules that form on the surface of the liquid. A liquid with a low surface tension is a better penetrant, thus having the ability to penetrate discontinuities.

> Wettability

It is the property that the liquid has of spreading on the surface, i.e. the better the wettability, the better the penetrant.

> Volatility

As medium viscosity is desirable and volatility is linked to viscosity, medium volatility is also desirable. The more volatile, the shorter the penetration time.

> Melting Point

This is the temperature at which the presence of a flame can ignite it. For a good penetrant, the flash point must be high, i.e. above 200°C.

The Revealer that will highlight and reveal the discontinuities has the following main characteristics:

> Absorb the penetrant from the discontinuity;

> Serve as a base for the penetrant to spread through;

> Must be easily removable

A corrosion inhibitor must be added to the solution and the concentration must be controlled, as evaporation can occur.

4.1.3 ULTRASOUND TESTING

Ultrasonic testing is based on the phenomenon of acoustic wave reflection, where the wave is sent out and then reflected back to its source.

According to ZOLIN, ultrasound is the most widely used non-destructive testing method worldwide for testing internal discontinuities in materials. Ultrasound is an acoustic wave with frequencies above the audible limit. Typically, ultrasonic frequencies are in the range of 0.5 to 25 MHz. The ultrasonic pulse is transmitted to the material via a special transducer, usually called a head. The ultrasonic pulses reflected by a discontinuity, or by the opposite surface of the workpiece (background echoes), are picked up by the transducer, converted into electronic signals and displayed on the device's flat liquid crystal display. Generally, the real dimensions of an internal discontinuity can be estimated with reasonable accuracy using the height of the reflected echoes, providing the means for the part to be accepted or rejected, based on the acceptance criteria of the applicable standard. The main applications of this test are in welds, laminates, forgings, castings, composite materials, thickness measurement, corrosion, etc. Ultrasonic testing is undoubtedly the most widely used and fastest growing non-destructive testing method for detecting internal discontinuities. This is due to the ease with which the test is carried out, low investment, speed of execution and high sensitivity. Ultrasound has long been used in approval testing, process control, aircraft inspection and in the nuclear, petrochemical and steel industries.

Ultrasonic techniques are basically divided into two: contact and non-contact (immersion) techniques. In the contact technique, the transducer is directly applied to the object using water, oil or other agents that serve as a coupling medium. In the non-contact technique, the transducer is manipulated at a certain distance from the test object, within a medium that can be water or light oil, which has the advantage of eliminating the influence of coupling variation. The choice of technique should be made taking into account sensitivity, geometric shape of the part, type and orientation of the discontinuity, simplicity of operation, speed required for inspection, etc. The contact technique is more commonly applied to large products and welded structures, while the immersion technique is used for testing large batches of small, identical parts using automated systems, especially in the automotive and aeronautics industries where high sensitivity is required.

The discontinuities become visible on the liquid crystal screen in the form of an echogram produced by an electronic signal (echo signal x time).

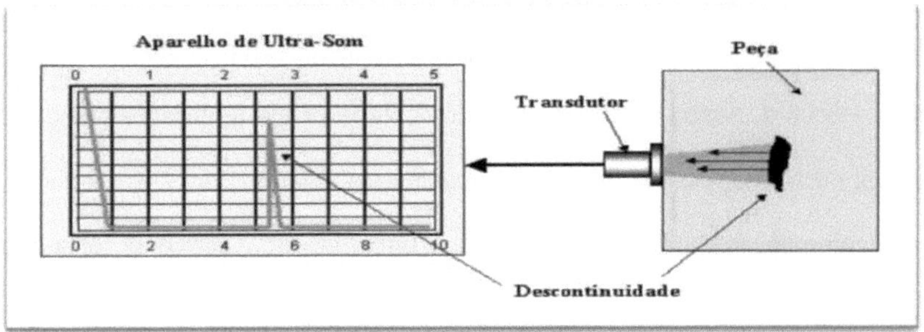

Figure 2 - Echogram with discontinuity (ANDREUUCCI, 2008)

5 METHODOLOGY

In this research, X-ray, Liquid Penetrant and Ultrasound tests were carried out on test specimens. The specimens used for this purpose were carbon steel plates and tubes after the welding process in order to detect discontinuities and certify the reliability of the product being manufactured. The specimens were manufactured by Estaleiro Vard Niterói S.A. and are used for the construction of Offshore Support Vessels. The procedures are in accordance with ASME Section V Non-Destructive Testing and will be described in each test.

5.1 SPECIMEN PREPARATION

The specimens used for the tests were made from carbon steel plates and tubes with different dimensions for each type of test, as shown below:

> X-ray testing

Two ASTM A106 Grade B carbon steel tubes with a thickness of 6.02 mm and a diameter of 4 inches were used for this test. The welding process used a V-notch butt joint and TIG *(Tungsten Inert Gas) welding*. The height of the weld bead after the welding process was set at 3 mm and its width at 10 mm. This test was carried out after the parts had been welded.

> Liquid Penetrant Testing

Two HELDOX 700 carbon steel plates 40 mm thick, 500 mm long and 250 mm wide with a V-shaped butt joint were used for this test. Preheating was carried out to remove moisture from the steel plate and then the MIG *(Metal Inert Gas)* welding process began, using carbon steel wire in a horizontal position. This test was carried out before and during the welding process.

> Ultrasound testing

Two HELDOX 700 carbon steel plates 40 mm thick, 500 mm long and 250 mm wide with a V-shaped butt joint were used for this test. Preheating was carried out to remove moisture from the steel plate and then the MIG *(Metal Inert Gas)* welding process began, using carbon steel wire in a horizontal position. The height of the weld bead after the established welding process is 5 mm and its width is 40 mm. This test was carried out after the entire welding process and the liquid penetrant test had been completed.

5.2 X-RAY TESTING

Procedures and performance of the Radiography (X-Ray) test, carried out by the SPERJ team following the procedures contained in the ASME Section V standard, articles 2 and 22, 2001 edition (Non-destructive examinations - Radiography) at the VARD NITERÓI S.A. shipyard in order to detect discontinuities after the welding procedure (ANDREUCCI, 2008).

Before subjecting the welded part to radiation, it is necessary to prepare the equipment that will be used. Below are the main stages of this preparation.

> Assemble the system in the bunker, which is where the X-ray tube and film are housed, placing the piece in the radiation beam of the device. The bunker can be made of a lead-coated metal structure or concrete walls for larger pieces.

> We will now close the bunker door, as this is the only way the equipment can be activated, due to the safety system, thus avoiding exposing operators and the environment to radiation.

> Adjust the equipment via its control panel, setting the exposure time, voltage
and amperage.

Figure 3 - Industrial X-ray machine (Maxim, 2015)

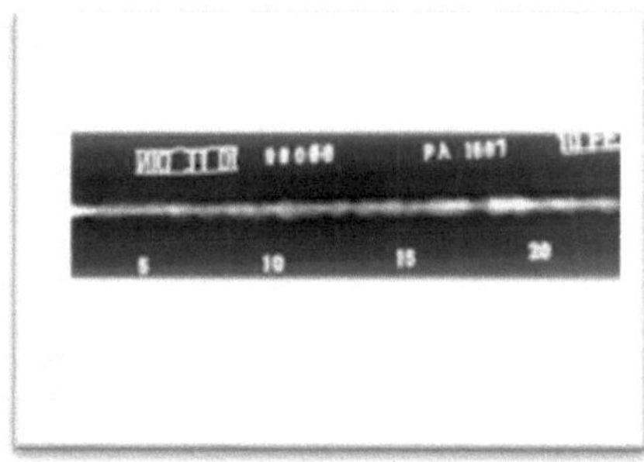

Figure 4 - X-ray film (Maxim, 2015)

After adjusting the device, we will now move on to the X-ray testing procedures.

> Activate the safety system to ensure that during the test the radiation levels
are within the safety standards required by the norm.

> Turn on the tube refrigeration and wait for the exposure time, which is
stipulated by a graph provided by the manufacturer depending on the thickness of

the material and the voltage used in the test.

> Once the exposure time has elapsed, the equipment is switched off and the film to be processed is removed from the bunker chamber.

> Then we move on to film development, which consists of chemical processing in order to reveal and fix the image obtained on the radiographic plate.

> To finish the process, we will now analyze the quality of the radiograph through a negatoscope for a better view. At this stage, we will be able to see the existing discontinuities according to the images below.

Figure 5 - Film after the test (SPERJ, 2015)

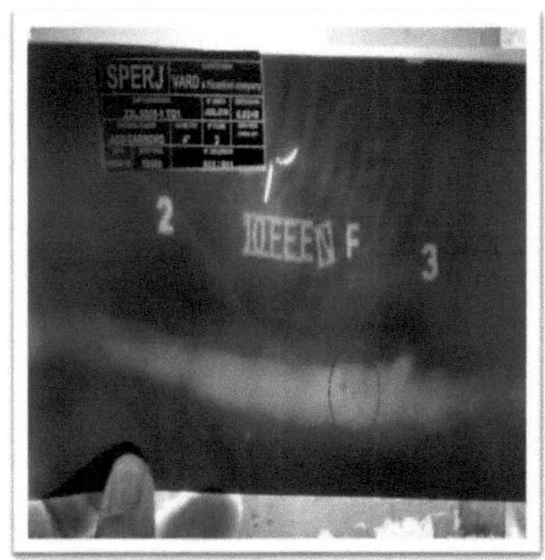

Figure 6 - Film after the test (SPERJ, 2015)

Figure 7 - Film after the test (SPERJ, 2015)

Figure 8 - Non-compliance report (SPERJ, 2015)

Through the images of the films obtained, we certified the presence of porosity acquired by the welding process of the material, which is why in this case the piece was rejected as shown in the non-conformity report submitted by the SPERJ team.

5.3 PENETRATING LIQUID TEST

Procedures and performance of the Liquid Penetrant (LP) test, carried out by the PW END team (2015) following the procedures contained in ASME standard Section V, 2004 edition (Non-destructive examinations - Liquid Penetrant) at the VARD NITERÓI S.A. shipyard in order to detect discontinuities after the welding procedure between plates (PUJUTTI, 2011).

> Surface preparation

The first step in the test is to clean the surface of the part, which must be free of residues, dirt, oil, grease or any other contaminant that could obstruct the discontinuities to be detected.

22

Figure 9 - Dirty surface material (PW END, 2015)

Figure 10 - Material with a clean surface (PW END, 2015)

> Application of the Penetrant

We must be careful with the surface temperature before applying the penetrant, where the ambient temperature must be close to 20°C and the surface temperature must not be below 10°C. These temperatures are extremely important for a good test result.

We apply a spray penetrant to the surface to be analyzed.

Figure 11 - Penetrant application (PW END, 2015)

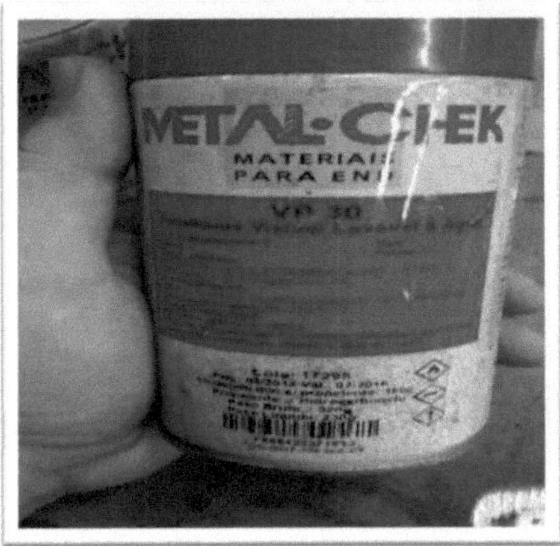

Figure 12 - Liquid penetrant spray (PW END, 2015)

> Penetration time

This is the time required for the penetrant to enter the discontinuities. This time depends on the material to be tested, the temperature and must be in accordance with the standard, in the case of the material in question the time is 15 minutes.

Figure 13 - Surface with penetrant (PW END, 2015)

> Removal of excess penetrant.

We will use water in this case, since the penetrant is water-washable, to remove the excess penetrant, and then dry the surface of the piece with a cloth.

Figure 14 - Removal of excess penetrant (PW END, 2015)

Figure 15 - Surface cleaning (PW END, 2015)

> Developer application

With the surface clean, we apply the Developer in the form of a spray to leave a thin, even layer. The developer dries for about 15 minutes until it is completely white and dry.

Figure 16 - Application of Developer (PW END, 2015)

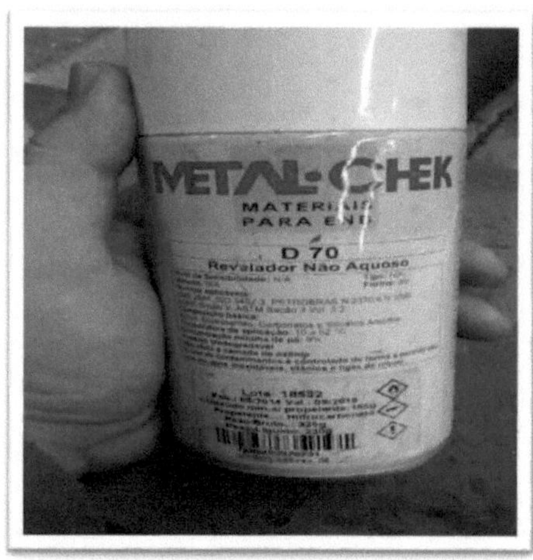

Figure 17 - Spray Developer (PW END, 2015)

> Drying and Inspection

As soon as drying begins, the evolution of the indications must be monitored in order to define and characterize the type of discontinuity. Fine, shallow discontinuities take longer to develop, unlike larger ones which quickly stain the developer.

Figure 18 - Developer drying (PW END, 2015)

Figure 19 - Revealing discontinuities (PW END, 2015)

> Evaluation and Acceptance Criteria

The discontinuity must be analyzed by the company's quality control standards, and if it fails it will constitute a defect, in the case shown above we have some red dots that show a discontinuity such as a crack or a pore, which is why it failed and had to remove part of the weld and redo the welding procedure and then repeat the Liquid Penetrant test again to certify that there was no discontinuity in this type of test.

5.4 ULTRASOUND TEST

Procedures and performance of the Ultrasound (US) test, carried out by the PW END team (2015) following the procedures contained in the standard used for this test was ASME Section V, edition 2010, (Non-destructive examinations - Ultrasound) at the VARD NITERÓI S.A. shipyard in order to detect discontinuities after the welding procedure between 40 mm thick HELDOX 700 carbon steel plates. This test was carried out after the liquid penetrant test in order to check for discontinuities that had not been detected by the previous test (ZOLIN, 2011).

> Scanning surface preparation

Weld inspection is carried out on the surface of the metal in an area that extends immediately from the weld bead, which we call the scanning area or surface, so we must remove paints, oxides, grease and anything else that could hinder or prevent the penetration of the sonic beam into the part to be tested. The temperature of the part cannot normally exceed 60°C, which can damage the transducers.

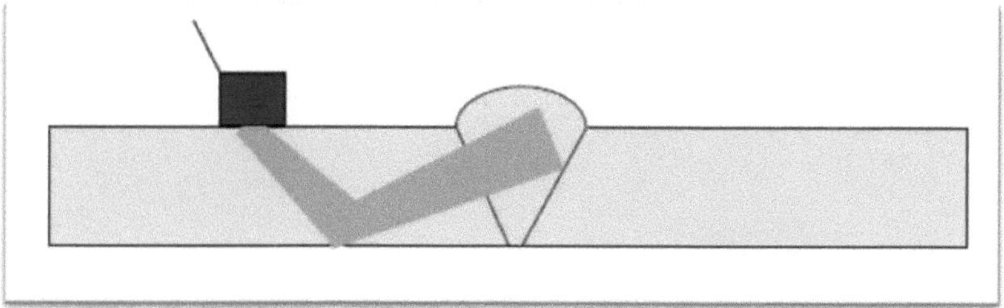

Figure 20 - Ultrasonic weld inspection technique (ANDREUUCCI, 2008)

> Device sensitivity calibration

The scale of the device must be calibrated using standard blocks, while the sensitivity must be calibrated using a block with calibrated thicknesses and reference holes and made of material that is acoustically similar to the part being tested. The material we are analyzing generally uses a frequency of 4 MHz and angles of 45 to 60 degrees.

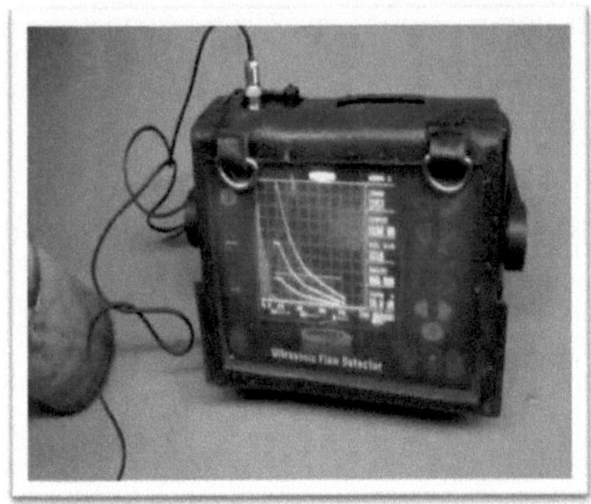

Figure 21 - Ultrasound device (PW END, 2015)

Figure 22 - Adjusting the device (PW END, 2015)

> Carrying out the inspection

To ensure the passage of the single beam, it is necessary to use a coupling liquid, such as methyl cellulose solution. The transducer must then be slid over the scanning surface with the ultrasonic beam facing perpendicular to the weld. If there is any discontinuity, there will be a reflection and this will indicate on the device's screen and a red light will flash, indicating that there is a discontinuity at this point and it will be analyzed whether it is acceptable or not.

Figure 23 - Application of the solution (PW END, 2015)

Figure 24 - Performing the test (PW END, 2015)

> Evaluation and Acceptance Criteria

The judgment of discontinuities found must be made in accordance with the procedure, applicable standard, customer specification or the company's own quality control.

In general, discontinuities are judged by their length and the amplitude of the reflection echo. Discontinuities must also be evaluated and are decisive in the acceptance or rejection of the welded joint. Discontinuities can be called cracks, lack of fusion or lack of penetration.

Figure 25 - Performing the test (PW END, 2015)

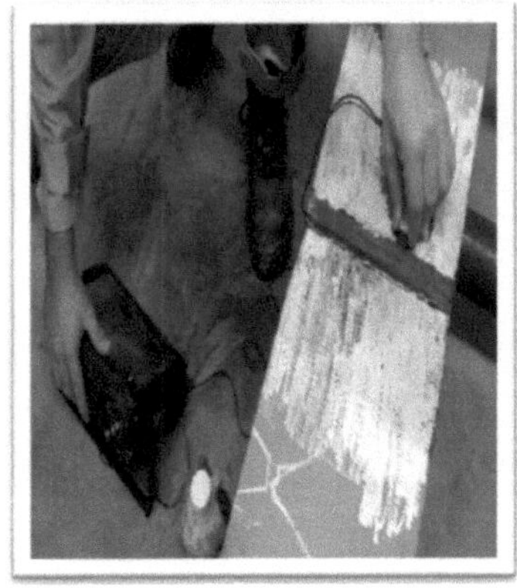

Figure 26 - Carrying out the test (PW END, 2015)

6 Results and discussion

Based on the tests carried out above, we can say that Non-Destructive Testing (NDT), besides being easy to apply, is an essential tool for defining the quality and reliability of products in their manufacturing phase.

But in order for the test to be valid and certified, it must be carried out by qualified professionals, following the appropriate procedures for both the equipment and the condition of the part to be analyzed, where cleaning the surface of the material is extremely important in order to obtain the best result.

The techniques used in this study are the most commonly used in industries to detect discontinuities and flaws in welded and cast parts. They are simple to apply and provide extremely satisfactory results in order to prove the reliability of the products and services provided. They are normally prepared and monitored by the Quality Control (QC) department of each industry and the reports are filed away to be delivered with the final product as proof of possible future problems.

6.1 x-ray scanning

The acceptance criteria for any component inspected by radiography must be assessed in accordance with the standard applied to the procedure, i.e. ASME Section VIII div.1, (Non-destructive testing - Radiography of welded joints). There is no possibility of evaluating radiographs on the basis of standards other than those applied to this test or even on the basis of the inspector's experience.

The welds must be free of any indications such as cracks, fusion zones or incomplete penetration, regardless of their length, otherwise the part will be rejected by the inspector in accordance with the established standards.

After the X-ray test, we verified the presence of porosity identified on the radiographic film, which caused a non-conformity according to the established standard, i.e. a pore was detected in the fusion zone during the inspection and was therefore immediately rejected. In order to solve the problem, it will be necessary to remove the weld in the area detected and in the surrounding area and to redo the welding process of the parts again in accordance with the established standard and, after completion, repeat the X-ray test to check that the discontinuity has been completely eliminated.

6.2 LIQUID PENETRANT TEST

The acceptance criteria for discontinuities must follow the standard applied in the procedure, i.e. ASME Section VIII Div.1 Appendix 8 which is the same as ASME Section VIII Div. 2 Art. 9-2 para. 9-230 (Non-destructive testing - Liquid Penetrant).

An indication is evidence of a mechanical imperfection, only indications larger than 1/16 inch (1.6 mm) should be considered relevant. A linear indication is one with a length greater than three times the width, a rounded indication is one in a circular shape with a length equal to or less than three times the width, or any questionable or doubtful indication must be re-inspected to determine whether or not relevant indications are present.

For acceptance criteria, all surfaces must be free of relevant linear indications and relevant rounded indications larger than 3/16" (4.8mm). An indication of an imperfection can be larger than the imperfection, however, the size of the indication is the basis for the acceptance assessment.

The Liquid Penetrant test is carried out at the beginning and during the welding process to detect discontinuities such as flaws and cracks. In the experiment carried out, the presence of a flaw was detected after the developer had dried, so a part of the weld where the detected discontinuity occurred was removed and the welding process was carried out again in accordance with the established standards and the test was carried out again to make sure that the flaw had been eliminated. This test is extremely simple and functional, making it widely used in industry.

6.3 ULTRASOUND TESTING

The judgment of the discontinuity found must be made in accordance with the same standard applied in the procedure, i.e. ASME Section VIII Div.1, Div. 2 and Section I. (Non-destructive testing - Ultrasound). In general, they are judged by the length and amplitude of the reflection echo, which are quantities measurable by the ultrasound inspector.

Indications characterized as cracks, lack of fusion or incomplete penetration, regardless of their length, must not be accepted by the inspector in accordance with the established standards.

In the Ultrasound test, which was carried out after the Liquid Penetrant test, a flaw was detected and subsequently corrected, so the Ultrasound test was carried out to make sure that the flaw in the welding process had really been eliminated, and after the test we made sure that there was no obstacle to the reflection of the wave, i.e. there was no flaw in the welding process of the parts. In accordance with the standard used, the part did not show any type of discontinuity such as cracks, lack of fusion or incomplete penetration, and in this case, the part was approved after being inspected and judged by the inspector. If we hadn't carried out the Liquid

Penetrant test during the welding process, we would certainly have had a fault at the end of the process after the Ultrasound test and we would have had to remove all the welds made at the junction of the parts until we found the fault and then redo the welding process and the Ultrasound test again in order to make sure that the discontinuity or fault had been eliminated.

7   CONCLUSION

We conclude that after carrying out the tests discussed above, they are of great value in the manufacturing process, as they show faults and discontinuities during the process with great precision, The x-ray and liquid penetrant tests in turn showed us a small discontinuity in the weld made at the junction of the parts, such as pores in the x-ray and red dots in the liquid penetrant, which may also be pores, and may have been caused by a small fault on the part of the welder or a fault caused by the adjustment of the machine or the gas. The ultrasound test, on the other hand, did not detect a fault because it had already been corrected in the liquid penetrant test in the middle of the welding process, and until its conclusion there were no further faults such as those of the operator or inadequate regulation of the machine or gas.

By carrying out some tests such as Liquid Penetrant, Ultrasound and X-Ray during this work, we gained some experience in using the equipment. We can conclude that penetrant liquid testing, despite being quite simple, is in many cases capable of detecting discontinuities that neither X-ray nor ultrasound can, and at a relatively lower cost. Ultrasound testing, on the other hand, is more complex and requires experience in detecting discontinuities. Ultrasound testing is currently one of the most widely used tests, with great potential for detecting internal flaws, and is the only test capable of analyzing large parts. X-ray testing also requires a certain level of experience both in image processing and in detecting discontinuities, which is very useful in carbon steel tubes.

In general, the study carried out in this paper shows that the NDT study carried out at the Vard Niterói S.A. shipyard was of great value in improving knowledge and analysis of materials and their processes.

8   ANNEXES

Annex 1 - X-ray test report

<table>
<tr><td colspan="3">SPERJ</td><td colspan="6">RADIOGRAFIA
RELATÓRIO PARA REGISTRO DE
RESULTADOS DE ENSAIO</td><td colspan="4">FOLHA : __ 1_/_1__
Nº SPERJ :076/14
Nº DO CLIENTE : _____
(OPCIONAL)</td></tr>
<tr><td colspan="3">CLIENTE:
VARD – a Fincantieri company</td><td colspan="10">OBJETO DO ENSAIO (Equipamento, Componente, Peça, Corpo de Prova):
REDE DE CARGA(GLP) EP 01</td></tr>
<tr><td colspan="3">MATERIAL:
A/C</td><td colspan="5">PROCEDIMENTO (REVISÃO)
PO 012 / REV 07</td><td colspan="5">NORMA DE INTERPRETAÇÃO
ASME B31.3 / ASME IX -2010</td></tr>
<tr><td colspan="3">FONTE RADIOATIVA - TIPO
Ir 192</td><td colspan="3">FONTE RADIOATIVA N SÉRIE
IRS- 6464</td><td colspan="3">TAMANHO FOCAL (mm)
3.0X2.0</td><td colspan="2">DFF (mm)
114/220mm</td><td colspan="2">FILME (FABRICANTE E TIPO ASTM)
KODAK M100 / AA400</td></tr>
</table>

IDENTIFICAÇÃO	JUNTA	POS	ESP. (mm)	SOLDADOR	POS DE SOLD	PROC DE SOLD	TÉCNICA RADIOGRÁFICA (a)	(b)	(c)	DESCONTINUIDADE	A	R	REC	DIÂM.
43L.7100.2	01	1	5.16+R	658 / 373	**	TIG/ER	X			PO	X			2.5"
43L.7100.2	01	2	5.16+R	658 / 373	**	TIG/ER	X			PO	X			2.5"
43L.7100.2	01	3	5.16+R	658 / 373	**	TIG/ER	X			PO		X		2.5"
43L.7100.2	01	4	5.16+R	658 / 373	**	TIG/ER	X			PO	X			2.5"
43L.7100.2	02	1	5.16+R	658 / 373	**	TIG/ER	X			PO		X		2.5"
43L.7100.2	02	2	5.16+R	658 / 373	**	TIG/ER	X			PO	X			2.5"
43L.7100.2	02	3	5.16+R	658 / 373	**	TIG/ER	X			PO		X		2.5"
43L.7100.2	02	4	5.16+R	658 / 373	**	TIG/ER	X			PO	X			2.5"
22L.0005.2	09	1	6.02+R	372 / 372	**	TIG/ER	X			*****	X			4"
22L.0005.2	09	2	6.02+R	372 / 372	**	TIG/ER	X			*****	X			4"
22L.0005.2	09	3	6.02+R	372 / 372	**	TIG/ER	X			*****	X			4"
22L.0005.2	09	4	6.02+R	372 / 372	**	TIG/ER	X			*****	X			4"
32L.7100.3	01nr	5	5.18+R	372 / 707	**	TIG/ER	X			*****	X			6"
32L.7100.3	02nr	5	5.18+R	372 / 707	**	TIG/ER	X			*****	X			6"
32L.6045-1	04nr	2	6.02+R	372 / 392	**	TIG/ER	X			FF		X		4"
32L.7400-8	01nr	4	5.49+R	372 / 707	**	TIG/ER	X			*****	X			3"
22L.7250-1	01nr	3	6.02+R	863 / 707	**	TIG/ER	X			*****	A			4"

OBSERVAÇÃO
PROGRAMAÇÃO : 042/14

SPERJ/076/14
n.o. 08/18/14

CLEBER DA SILVA LIMA
SNQC .0816-SR-N2-RT
Nível II 120/00
Nível 1 (Opcional)

CLIENTE: 15/10/14

Annex 2 - Liquid penetrant test report

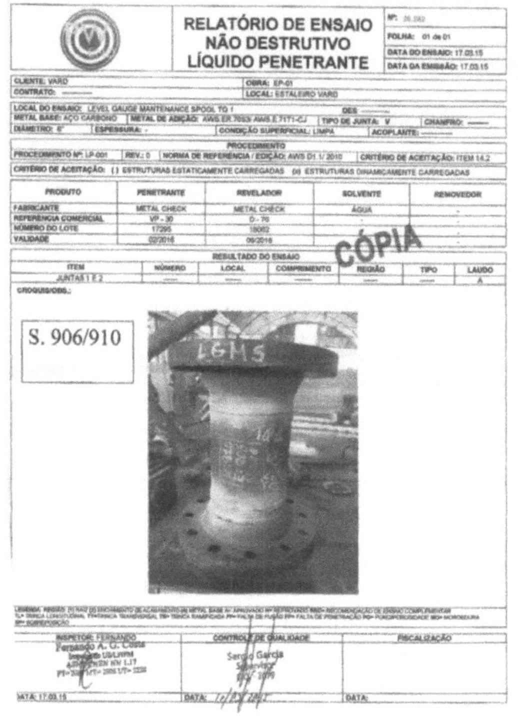

Annex 3 - Ultrasound test report

RELATÓRIO DE ENSAIO NÃO DESTRUTIVO
ULTRASSOM

Nº: 1827 FOLHA: 01 de 01 DATA: 08.09.2015

CLIENTE: VARD
CONTRATO: —
OBRA: PRO 32
LOCAL: ESTALEIRO VARD

LOCAL DO ENSAIO: EMENDA DAS UNIDADES 266 X 28P
METAL BASE: AÇO CARBONO   METAL DE ADIÇÃO: AWS E-7111   TIPO DE JUNTA: -   CHANFRO: -
DIÂMETRO: -   ESPESSURA: -   CONDIÇÃO SUPERFICIAL: ESCOVADA   ACOPLANTE: METIL CELULOSE

PROCEDIMENTO

PROCEDIMENTO Nº: US 002   REV.: 05   NORMA DE REFERÊNCIA / EDIÇÃO: AWS D1.1 M : 2015   CRITÉRIO DE ACEITAÇÃO: ITEM 15

CRITÉRIO DE ACEITAÇÃO: ( ) ESTRUTURAS ESTATICAMENTE CARREGADAS   (x) ESTRUTURAS DINAMICAMENTE CARREGADAS

EQUIPAMENTO

APARELHO: DIGITAL   FABRICANTE: MODSONIC   MODELO: EINSTEIN-II DGS   NÚMERO DE SÉRIE: EX139-0811
BLOCO PADRÃO: V1   Nº CERTIFICADO: 10521/14   REGISTRO DE CALIBRAÇÃO: 10521/14

CABEÇOTES UTILIZADOS

TIPO	FABRICANTE	MODELO	NÚMERO	DIMENSÃO (mm)	ÂNG. REAL	FREQ. (MHz)	GR	GV
DUPLO CRISTAL	MTECH	MSEB	130456	Ø 10	-	4	58.0	66.0
ANGULAR	MODSONIC	SW 70 NC	MODSONIC	20x22	70°	2	58.6	70.6

RESULTADO DO ENSAIO

IDENTIFICAÇÃO	Nº INDICAÇÃO	CABEÇOTE	SUPERFÍCIE	PERNA	GANHO DA INDICAÇÃO (A)	GANHO DE REFERÊNCIA (B)	FATOR DE ATENUAÇÃO (C)	CLASSE DA INDICAÇÃO (D)	COMPRIMENTO	PERCURSO SÔNICO	PROFUNDIDADE	DISTÂN. CIR. A PARTIR DE "S"	A PARTIR DE "T"	LAUDO	GEOMETRIA / REGIÃO	TIPO DE DEFEITO	SOLDADOR
PONTO 1	-	-	-	-	-	-	-	-	-	-	-	-	-	A	-	-	10-40
PONTO 2	-	-	-	-	-	-	-	-	-	-	-	-	-	A	-	-	10-40

CROQUIS/OBS.:

LEGENDA:
A-Aprovado
R-Reprovado
ABS-Absolutamente
Pouco Considerável

GRAU DE DIFICULDADE
GEOMETRIA DA JUNTA
1-Junta de Topo
2-Junta de Ângulo
3-Junta em "T"
4-Junta de Canto

SEÇÃO DA JUNTA
1-Raiz
2-Enchimento
3-Acabamento
4-Metal de Base

TIPO DE DEFEITO
1-Trinca
2-Poro
3-Falta de Fusão
4-Falta de Penetração
5-Escórias de Polarização

4-Porosidade
5-Mordedura na Raiz
6-Cadinho de Escória
7-Inclusão Tungstênio

INSPETOR: FERNANDO
Fernando A. G. Costa
ASNT-SNT NÍVEL II

CONTROLE DE QUALIDADE

FISCALIZAÇÃO

DATA: 08.09.2015   DATA: DATA:

9 BIBLIOGRAPHICAL REFERENCES

- American Society of Mechanical Engineers (ASME). **Boiler and Pressure Vessel Code.** Section VIII Div.1 and 2.

- Krautkramer. **Ultrasonic Testing of Materials**. Germany, Second Edition.

- LEITE, Paulo G.P. **Course in Non-Destructive Testing**. Brazilian Metals Association (ABM). Sao Paulo, 1966, 8^a ed.

- PETROBRAS. **Standard N-1595 d**, Non-Destructive Testing - Radiography.

- National Learning Service (SENAI). **Welding.** Sao Paulo, 1997.

- ANDREUCCI, Ricardo. **Industrial Radiology**. Ed. Jan./ 2008.

- ANDREUCCI, Ricardo. **Ultrasound testing**. Ed. Jul./ 2008.

- ANDREUCCI, Ricardo. **Penetrating Liquids**. Ed. July/ 2003.

- ZOLIN, Ivan. **Mechanical testing and failure analysis**. Ed. 2011.

- PUJUTTI, Fabricio. **Mechanical Maintenance - Maintenance Techniques** - Ed. 2011.

Web pages:

- Association of Inspectors. Available at:

http://blog.associacaodeinspetores.com.br/. Accessed on: June 1, 2015.

- Compoende company website. Available at:

http://www.compoende.com.br/index.php. Accessed on: June 1, 2015.

- Info Solda website. Available at: http://www.infosolda.com.br/. Accessed on: June 1, 2015.

- http://www.abendi.org.br/ - Brazilian Association of Non-Destructive Testing and Inspection.

More Books!

yes

I want morebooks!

Buy your books fast and straightforward online - at one of world's fastest growing online book stores! Environmentally sound due to Print-on-Demand technologies.

Buy your books online at
www.morebooks.shop

Kaufen Sie Ihre Bücher schnell und unkompliziert online – auf einer der am schnellsten wachsenden Buchhandelsplattformen weltweit! Dank Print-On-Demand umwelt- und ressourcenschonend produziert.

Bücher schneller online kaufen
www.morebooks.shop

info@omniscriptum.com
www.omniscriptum.com

OMNIScriptum

Printed by Books on Demand GmbH, Norderstedt / Germany